£2

CW01663872

"They Don't Pass Anyone on Friday Afternoons!"

Written and illustrated by

© Bob Harrod (Dtp ADI)

1st Edition, September 1989
Revised December 1989
Revised May 1990
Revised April 1991

Published by:-
Iron Box Publications
Iron Box Cottage, Whitecroft,
Dilton Marsh, Westbury
Wiltshire BA13 4DJ

Typeset and printed by:-
Kall Kwik Printing, Salisbury, Wiltshire

Preface

Or do they? The truth is, 'they', being the Department of Transport's vast army of driving examiners, are all fairly anxious to knock off by 5pm every day (Monday to Friday) and will pass anyone who can 'drive' around a fairly simple test route for about thirty minutes, without committing either a *serious* or *dangerous* error. For some reason known only to the higher-up officials in the Department of Transport, it is not even necessary for a test candidate to know the Highway Code. Although it must be said that perhaps if 'they' failed a few people on their lack of knowledge of this subject we would not have to bury in excess of 5,000 victims of ignorance and blind stupidity every year. However, this book has not been written with the intention of passing judgement on anyone.

The primary objectives of this book (other than to make a pile of money so that I do not have to give so many driving lessons in order to feed me and my family) are threefold: firstly to explode a couple of the popular myths about driving tests, and driving examiners, e.g. "they don't pass anyone on Friday afternoons!" Secondly to try and advise the public about what to look for (and what to avoid) when trying to choose a driving school to give their hard earned cash to. Finally I have attempted to translate the Driving Test Statement of Failure (DL 24) into English. This of course was the most difficult of my tasks, as (a) it is the largest part of the book, and (b) the DL 24 was invented by civil servants.

This book will annoy many people, because they will recognise themselves when they read it. It will amuse many people (I hope!) but most important of all, it will throw a great deal of light on to what has for years been a very dark subject surrounded by myths and plain good old fashioned lies. I hope this book will be useful to people who are thinking of starting driving lessons, as well as those that have recently failed a driving test. It could also be useful to a few driving instructors who perhaps do not know as much as they should.

I have been a driving instructor for seventeen years, but this book is not written purely from my own experiences and knowledge, I owe thanks to my good friend Ken Simpson, Dtp ADI, (ex Department of Transport Driving Examiner) who filled in a lot of the grey areas for me, the details of which are only known to 'them'.

"At last, someone has got it right!"

"This book contains a mine of really useful information, about the Driving Test, the Examiners, and other drivers. It is written in a manner which makes it very difficult indeed to put down, it is very funny, and that alone makes you want to go on reading it in order to get to the next little gem, but in doing so you will be absorbing a large amount of very real information, which certainly makes it, in my opinion far and away the best book of its kind that I have read.

It is not for the learner alone, all drivers will find something useful in it, indeed some of them will see themselves hidden in its pages, and even if you have no intention of ever learning to drive, its humour, and the many amusing drawings it contains, make it easy reading indeed. But more important is the excellence of the advice and help it will give to anyone who reads it, from driver to pedestrian, I thoroughly recommend it.

I only hope that I never looked like the examiner in the drawings — although I have an awful feeling that I probably did!"

Ken Simpson
(Ex-Department of Transport Driving Examiner)

INDEX

Myths

There are, and probably always have been, a number of myths surrounding the driving test and driving examiners. Once in a while a new one crops up, but generally they are the same old stories occasionally given a slightly different flavour to suit the story teller's use in his or her excuse for failing the driving test. The Department of Transport must itself bear some of the responsibility for these fictitious stories circulating, due to the veil of secrecy they insist on drawing over the whole performance. The majority of failed candidates are left staring at an official looking piece of paper that usually means absolutely nothing to them. Quite often because when the offending mistake was made the candidate was not even aware of it. It is not widely known, and the Department of Transport is happy to keep it that way, that the candidate has the right to request a written report on the driving test from the area traffic office where the test was booked. This report will arrive in letter form and will state exactly and clearly what the candidate did to justify a fail. The examiners do now give a short verbal explanation, but the candidate, understandably being a little upset, will often misinterpret the examiners remarks, whose advice is then useless, and sometimes leads to discussion and even confrontation. So the system as it is does not solve the problem because of the mystery it leaves, so sometimes when a confused candidate has failed on a Friday afternoon, he will say "they never pass anyone on Friday afternoons!" This is because he has been told by his mates down the local the age old story of examiners having 'quotas' and that they must keep the national pass rate down to around 40%. Of course by Friday afternoon they have passed all the candidates for that week that they can afford to! The truth is, as any examiner will tell you, with the standard among driving test candidates being as low as it is, the difficulty is in trying to get the national average *up* around 40%. Keeping it *down* is no problem at all, the public manage that quite easily on their own.

Before we leave the subject of "quotas", I must relate a story which was told to me by one of my pupils a few years ago. It has probably done the rounds, but I have only heard it told seriously on the one occasion. A young lad had just passed, and I was driving him home. As we were going along, he said, "my dad took two tests you know!" I said "fancy that!" or some other stupid remark, trying to sound surprised.

He went on to say that after the first test, the examiner said "well Mr Bloggs, you drove perfectly, but unfortunately I've passed my quota for this week, so we will have to go around again so that I can find something wrong!" It took me a while to decide whether the father was the more stupid of the two for telling the story in the first place, or the son for believing it. I came to the conclusion that they must both be as daft as each other to think that someone could actually manage to drive *perfectly* in the first place.

One of the favourite stories that I personally have been told a number of times by failed candidates, neighbours and certain people who seem to find it necessary to exercise their voice boxes when they find themselves in the company of a driving instructor, is the one about the driving examiner, who, having settled himself in the passenger seat, suddenly remembers something that he has left in the office. (No-one has ever told me what, he probably forgot to put the cat out). The examiner dashes off back to the office, leaving his papers on the seat, and the candidate, unable to control his natural curiosity has a quick look through them and finds something that tells him that he has already failed. No doubt a victim of a fulfilled quota! I am unable to make any excuses for the people who tell this story, it is simply a way of explaining why they failed. It sounds better than "I did it wrong!"

One small point that perhaps should be mentioned, and this is just an observation, I draw no conclusions from it; every time an examiner fails a candidate he has to make out a written report (quite separate to the one which is given to the candidate) which is then forwarded to the traffic area office. These reports must be written by the end of that working day. No report is written when a candidate passes.

How to Choose a Driving School

This is a difficult question to answer. First of all I must point out that you are not buying a new pair of knickers or a loaf of bread, so you will not be able to walk round the corner to a different shop and buy the same for exactly the same money. I would be very surprised if there were two driving instructors in this country who were both exactly the same, and I bet if you were lucky enough to find two they would not have exactly the same car. What you need to find is an instructor who suits you and gives value for money. As far as suiting you goes, what I mean is, someone you feel happy with. The best driving instructor in the world may not suit you if there is a personality clash. As for getting value for money, if you don't pay much you may not get much, so what you do get will probably be the appropriate value for the money you pay. On the other hand if you choose an instructor who charges a little over the normal rate for your area, then you will also get value for money. However, paying more money, you will probably get better value. Always remember, most driving instructors know what they are worth and charge accordingly. There is not an instructor on this planet (or any other as far as I know) who earns less per hour than his counterparts because he likes it that way. The reason that some people charge less than others is simply because they doubt their ability to command the higher rate.

Why do some instructors doubt their ability? Well, first of all, to be fair to the newly qualified, an instructor who has recently set up on his own will almost certainly charge a little under the average rate for the area to attract new business. There is nothing at all wrong with that, we all did it. The problem for you, the customer, is that you have no way of telling whether or not the instructor in question is a good one or not. Time will tell, after the new instructor has been around a while he will either be busy or not. If he is, his hourly rate will rise to reflect his popularity, if he is not, his rate will stay down to attract business. This process of course will all take too long for you, you want to get on and pass your test. If you are unsure of an instructor then the best advice perhaps would be to avoid him and find one that you feel more sure about. This may not seem very fair to the new instructors, but when I think back to some of the things I said and did as a young newly qualified instructor, it is probably the best advice I can give.

To return now to the question of instructors doubting their own ability. Let's deal with the instructor who has been qualified a number of years, yet still charges very little for his services compared to other schools in the area. I will not try to be diplomatic about this, these people are to be avoided! As I said earlier, no instructor earns less than others because he likes it that way, there is always a reason for it. I can think of a number of examples of young learner drivers receiving less than satisfactory service from cheap instructors, but I will only mention one. Recently I was at the local driving test centre with a young candidate and a number of other instructors with their candidates. A young lady, who was in the company of an instructor I know to be inexpensive, went out with her examiner, who, two minutes later returned to collect the tape measure. This happens when the candidate has difficulty reading the number plate pointed out by the examiner. The normal procedure then is for the examiner to measure out the prescribed distance accurately with a tape measure and test the candidate again. If the candidate is still unable to read the plate, the test goes no further and the fee is lost. Another procedure then swings into action which I will go into in more detail later in the section of this book relating to the driving test. However, to return to this unfortunate young lady who was under the impression that she was receiving 'cheap' lessons. She failed the eyesight test and was not permitted to go any further. A failure form was issued and away she went in tears with her instructor. The question is, did she really get cheap lessons? One driving test fee lost, God knows how many lessons preparing for the test all while completely oblivious to the fact that her eyesight was below standard. If she had chosen an instructor who charged a decent rate he would have probably had a more professional approach and the situation would not have arisen. Checking the eyesight of a new pupil should be one of the first things an instructor does before the learner gets behind the wheel. Remember, you get what you pay for. One final note on the subject of cheap driving instructors. We all know about the cost of living, you do not need me to tell you how difficult it can be to keep the bills paid and your head above water. You will be much better off with an instructor who has a decent standard of living and does not work all the hours God sends, than one who has to burn the midnight oil taking pupils out at the odd sort of hours that *they* demand.

A subject we must cover is the difference between qualified driving instructors and trainee instructors. Most members of the public do not know that trainee instructors exist, the fact is they do, and they have to learn their trade somehow. The question is, given the choice, do you want them to learn their trade with you as a guinea pig? Having said that, a young keen fresh trainee instructor might be more use than an old tired qualified man who can barely drag himself out of bed in the mornings. (I've been looking in the mirror again!) Things have changed in recent years, time was when a trainee instructor could take learner drivers out with little or no training himself at all. I know, I did it. These days a trainee instructor is not permitted to give tuition for money until he has passed the first two parts of the qualifying examination. These are, first, the written test, and second an actual driving test. Only the third part has not yet been taken, and that is the practical test of instructional ability, although it must be said that at least a good proportion of the theory of this subject has been tested during the written examination. You will not find it difficult to recognise either an approved driving instructor or trainee. They both have to display their certificates in the windscreen of their tuition vehicles and this is what they look like:

Approved (qualified) instructor.
(Certificate seen from inside vehicle)
(GREEN)

Trainee Instructor Certificate
is similar but PINK

Certificates as seen from OUTSIDE vehicle

(GREEN)

Approved (qualified) Instructor)

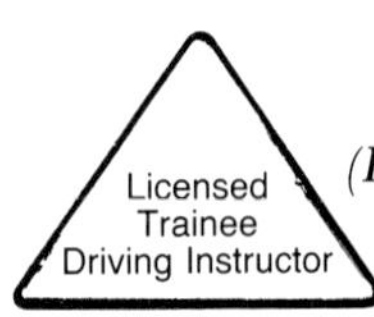

(PINK)

Licensed (trainee) Instructor

The obvious difference, without having to get too close, is the colour. Green for an approved instructor and pink for a trainee. Incidentally, it you do not see either a green or pink certificate in the windscreen of the vehicle, the 'instructor' is at best committing an offence by not displaying it, or at worst he does not have one and he is teaching completely illegally. Any money you pay one of these pirates will be *too* much, as their knowledge is extremely limited.

"Heave to on the left please, I think a police car is about to come alongside!"

Having dispensed with the instructors to be avoided, how do you actually choose one? There are only a few things you need to bear in mind. As I have said before, the better instructors charge a higher rate than normal. Finally, if you ask your friends and relatives who they would recommend, certain names will keep cropping up. All you have to do then is decide which sort of car you would prefer.

Try compiling a short list, giving them a ring and asking a few questions. You will soon know whether or not the person on the other end of the line is the sort of person whose company you could stand. But whatever you do, don't ask for discounts for block bookings, reductions because you are a nurse/OAP/unemployed or free gifts. You will be talking to someone who is very busy, who will not be amused and will more than likely tell you to naff off!

If you should experience difficulty in choosing a driving school or instructor, and you are tempted to give your custom to one of the large driving schools who perhaps have an office in your town centre, you may find that you are encouraged to pay for lessons in advance. Unless you wish to pay in advance you would be well advised not to. Most driving schools are extremely professional in their approach and will only have your best interests at heart. Unfortunately this cannot be said about all of them, and if after one or two lessons you are not receiving the sort of treatment you had expected to receive from someone who was receiving your money, you might be reluctant to change driving schools mid-stream if you had paid 'up front'. I am not trying to suggest that a refund would be refused, but you may feel awkward about asking. If you have not paid in advance you will not be obliged to say anything - except "don't ring me, I'll ring you".

The Driving Test Statement of Failure (DL 24)

The DL 24, or Statement of Failure is the piece of paper handed to unsuccessful candidates at the end of their driving test. A copy of this document is reproduced in this chapter, and I will go through the items in numerical order and attempt to explain each item in a little more detail.

Before we move on to the DL 24, just a few points regarding the procedure followed prior to actually driving off. First of all, when the driving examiner enters the driving test centre waiting room he will call out your name and will ask you to sign against your name on his pad. He will then ask to see your driving licence. If you do not have your driving licence for some reason, you **MUST** take some other form of identification with you, something preferably with a photograph like a passport or an I.D. card. **IF YOU ARE UNABLE TO PROVE YOUR IDENTITY, YOU WILL NOT BE PERMITTED TO TAKE THE TEST AND YOU WILL FORFEIT THE TEST FEE.** The other document you should have with you is your driving test appointment card which you will receive from the area traffic office after you have booked your test. Nine times out of ten, this card will not be needed, but mistakes do happen. I have been in our local test centre waiting room on a number of occasions when a certain candidate's name is not on the examiners programme although the candidate has been sent an appointment card. When this happens, the examiner will *always* ask to see the appointment card. I am not trying to say that you will automatically not be permitted to take the test if you do not have your appointment card, but it can only complicate matters. Bear in mind that the examiners work to a very tight schedule and the more time is wasted, the less chance you will have of taking the test that day.

The examiner will ask you to lead the way to your car, and you will leave the waiting room together. As you are walking to the car he will ask you if you have any disabilities that were not declared on your driving test application form. He will probably then talk about the weather. The whole idea is to relax you and not mention anything about driving as if the subject does not even exist. You and I both know that you are going to be thinking of nothing but driving, and by this time you will probably have convinced yourself that if you ever did know anything about the subject, you are totally unable to remember what it was. At this point you will probably even start to wonder if you can remember where you left the car! Which reminds me, one day I was outside our test centre waiting room when a candidate and his examiner walked out and got into one of two cars which were parked outside. One was bright red, the other bright yellow. After the candidate had tried for a few minutes to get the key in the ignition, both doors flew open and the occupants transferred themselves to the other car! So don't worry, the examiner will expect you to be nervous, even to the point of not being able to tell red from yellow. Hopefully though, by the time you encounter your first set of traffic lights during the test you will have sorted out a problem like that!

To return to the conversation about the weather, try to bear with the poor man, he probably is not any more interested in the weather than you are. Always remember, they are human too. They get divorced, have overdrafts and kick the cat when they get home just like lots of other people,(perhaps more than most!)

At this point you may get a shock that you were not expecting. It does not happen very often, but don't be surprised if a man dressed something like your examiner clutching a clipboard leaps out of the bushes grinning at you. No, not "This is Your Life" – he will be what is known as a supervising examiner. Your examiner will introduce him and will tell you that he is going to accompany you both on the test in the back seat of the car. He will then go on to say that you should not worry as the supervising examiner is not there to watch your driving or take any part in the test, he is only there to ensure that the test is conducted correctly, so you should disregard his presence. This advice probably has as much practical use as talking about the weather when all you can think about is driving! That is of course if like most people you have some sort of central nervous system. It may also be a bit difficult to disregard his presence in the back seat if he is 7 feet tall and weighs 22 stone and you are driving a mini!

"This is Mr Small, he will be coming
with us on the test, but I want
you to forget that he is there!"

We will now move on to the DL 24 in detail, but we will discuss the various points in the order that they appear on the form and not necessarily the order in which they would happen on the test.

Department of Transport

ROAD TRAFFIC ACT 1972

Test Centre:

Statement of Failure to Pass Test of Competence to Drive

Name ....

has this day been examined and has failed to pass the test of competence to drive prescribed for the purposes of section 85 of the Road Traffic Act 1972.

Date ....

Examiners have regard to the items listed below in deciding whether a candidate is competent to drive. The matters needing special attention are marked for your information and assistance and should be studied in detail. (See Note 1 overleaf)

1. ☐ Comply with the requirements of the eyesight test.
2. ☐ Know the Highway Code.
3. ☐ Take proper precautions before starting the engine.
4. ☐ Make proper use of/accelerator/clutch/gears/footbrake/handbrake/steering.
5. ☐ Move away/safely/under control.
6. ☐ Stop the vehicle in an emergency/promptly/under control/making proper use of front brake.
7. ☐ Reverse into a limited opening either to the right or left/under control/with due regard for other road users.
8. ☐ Turn round by means of forward and reverse gears/under control/with due regard for other road users.
9. ☐ Make effective use of mirror(s) well before / Take effective rear observation well before } signalling/changing direction/slowing down or stopping.
10. ☐ Give signals/where necessary/correctly/in good time.
11. ☐ Take prompt and appropriate action on all/traffic signs/road markings/traffic lights/signals given by traffic controllers/other road users.
12. ☐ Exercise proper care in the use of speed.
13. ☐ Make progress by/driving at a speed appropriate to the road and traffic conditions/avoiding undue hesitancy.
14. ☐ Act properly at road junctions:—
    —regulate speed correctly on approach;
    —take effective observation before emerging;
    —position the vehicle correctly/before turning right/before turning left;
    —avoid cutting right hand corners.
15. ☐ Overtake/meet/cross the path of/other vehicles safely.
16. ☐ Position the vehicle correctly during normal driving.
17. ☐ Allow adequate clearance to stationary vehicles.
18. ☐ Take appropriate action at pedestrian crossings.
19. ☐ Select a safe position for normal stops.
20. ☐ Show awareness and anticipation of the actions of/pedestrians/cyclists/drivers.

DRIVING EXAMINERS ARE NOT PERMITTED TO DISCUSS DETAILS OF THE TEST

SEE GUIDANCE NOTES OVERLEAF

DL24
9/81

1. **Comply With The Requirements of the Eyesight Test**

First of all what are the requirements of the eyesight test? You must be able to read a number plate at 67 feet. If you are asked to read an old type plate with the larger letters the prescribed distance is 75 feet. The examiner will first ask you to read a plate on a car which is obviously well outside the prescribed distance. If you are able to read it clearly, he will ask you to get into your car and the eyesight test will be over. If you have difficulty he will ask you to read a different plate on a different car. It may be that the first one was dirty, or the light was wrong for you or quite simply you were reading the plate on a car other than the one that the examiner intended you to. This can quite often happen as the examiner will not mention makes of cars as he can not expect you to know what the various makes look like, and will not mention colours because you may be colour blind. If there is an obvious problem, the examiner will fetch the tape measure from the office and will measure out the prescribed distance accurately.
He will then ask you to stand exactly 67 feet from the plate and ask you to try again. If you are still unable to read it you will return to the waiting room with the examiner. He will then issue a failure certificate and will ask you to sign a certificate which says that you agree that you have failed the eyesight test. The same rules apply then that would apply if you had failed the test while driving. In other words you will have to apply and pay for another test and will not be permitted to take it until one calendar month has elapsed.

The certificate that you sign is sent to DVLC, Swansea. They will then write to you and ask you to send an opticians certificate stating that your sight has been checked. You will then be permitted to re-take the test. If you fail your eyesight test a second time, DVLC will immediately revoke your driving licence and you will not be permitted to drive at all. This does not mean that you will never be able to have your licence back, but you will have to prove that your eyesight is up to the required standard before it will be re-issued.

All this time, trouble and expense can be avoided by making sure that your eyesight is up to scratch *before* you start driving. If you take lessons with a driving instructor it should be one of the first things he does before he lets you behind the wheel. Having said that, if he does not check your eyesight, remind him. He may simply have forgotten. On the other hand he may be one of the less efficient instructors who never bother anyway. At the end of the day, it is your eyesight, and you will be the one who is inconvenienced, not your instructor. If you drive while your eyesight is below the required standard you will be committing an offence, so it could cost you more than a wasted driving test fee. You might even kill someone!

Comply with the requirements of the eyesight test.

2. **Know The Highway Code**

I have yet to meet a learner who does! I always ask my pupils a few questions on the Highway Code just prior to their driving test, and the answers I receive range from comical to unbelievable. Most driving test candidates do not make a very serious attempt to learn this subject, others do not try as hard as they could and the rest do not bother at all. Most could not care less that the Highway Code is a complete mystery to them, but more to the point, the driving examiners do not seem to care either.

The blame for this complacent attitude towards road safety must rest mainly with the 'drivers' in this country who do not have the sense of responsibility to make themselves familiar with the rules regarding the use of the road. A good part of the blame however, must be laid at the door of the Department of Transport. The official line is that driving examiners *are* permitted to fail candidates on the Highway Code only, even if they have passed on their driving. However, I have never in the last fifteen years known anyone to fail *only* on the Highway Code. I have seen it on the failure form sometimes included with other faults, but I have *never* seen it on its own. I have sat in the back on driving tests and heard candidates answer *every* question incorrectly, and still pass. I often wonder if the examiners were to start failing people on their lack of knowledge, whether word would get round, and perhaps learners would make more effort. One thing is sure, word has definitely got round somehow that examiners are not *allowed* to fail people just on the Highway Code. This is totally untrue, but the public can be forgiven for believing it. Perhaps if the day ever comes when it will be necessary to pass a written test on the Highway Code before a provisional licence is issued, the situation may change.

Having said all this, don't made the mistake of thinking that you will pass whether you know the Highway Code or not. As I said before, I have never known anyone to fail *just* on the Highway Code, but I wonder how many there have been who failed on the driving part of their test because something happened as a direct result of their lack of knowledge? For example driving along at 30 miles per hour after passing a national speed limit sign either because they thought it meant something else or did not even see it in the first place.

Yes, the system is very slack, too many people are driving about with very limited knowledge. But if you want the best crack possible at the test, don't leave anything to chance. Read the Highway Code and make sure that you understand it. Buy one of the videos on the market, that will help.

Know the highway code.

3. **Take Proper Precautions Before Starting The Engine**

The question is, what are proper precautions? The text book answer of course is to check that the handbrake is on, and the gear lever is in neutral. But why? Quite simply because if the handbrake is off, and the gear lever is in first, when you turn the key the car will leap forward like a demented grasshopper and flatten any poor old lady who may have the misfortune of choosing to cross the road in front of your car.

Of course something else would have prevented this regrettable chain of events — putting the clutch pedal down of course! A proper precaution before starting the engine then could be depressing the clutch pedal.

Don't make the mistake of thinking that the examiner is sat there waiting for you to do something wrong, no matter how minor, so that he can pounce and fail you. He has to decide whether mistakes you make during the test are serious, dangerous or minor. You will only fail on serious or dangerous mistakes, minor ones will be marked but will not fail you. Not checking that the handbrake is on and the gear lever is in neutral will not even attract a minor mark. The examiner has to assume that you knew it was o.k. Not checking and having the car leap forward will probably only attract a minor mark, *unless it affects someone else*.
For instance, a pedestrian crossing the road or another vehicle in front of you. Say for example you stall at traffic lights, this alone will not fail you, but what action you take then might. There is no need to apply the handbrake and select neutral before starting the engine. All this performance takes ages in modern traffic conditions — you could even fail for being unduly hesitant if it took you too long! All you have to do is to depress the clutch and turn the key. However, if you forget to depress the clutch and the car leaps forward too close to the vehicle in front, certainly that could be a dangerous fault.

Remember, generally speaking a fault is only serious or dangerous as far as the examiners are concerned if it affects someone else.

Take proper precautions before starting the engine.

4. **Make proper use of/accelerator/clutch/gears/footbrake/handbrake/steering**

The items in this section that you could fail on are all to do with the actual control of the car rather than road procedure. If you were to fail for the incorrect use of any one of these controls then you probably should not be taking the test, because you have not yet mastered their use. No driving instructor can actually *teach* you how to be proficient in the use of the controls, he can only show you how to operate them. Proficiency will only come with practice, and that will cost you money. So if you object to paying for perhaps in excess of 25 lessons, even if you are fairly average, it will be a waste of time even starting lessons.

We will deal with the items individually starting with the accelerator. I have known candidates fail on this point in the past, and almost without exception they think that the examiner is referring to their speed.

What they usually say is "but I didn't go over 30 throughout the whole test!" The thought of someone tearing up to a pedestrian crossing full of people at 30 terrifies me, but that is another matter. A mark under this section has nothing to do with going too fast, that would be marked under item 12 (Exercise proper care in the use of speed) or 11 (Take prompt and appropriate action on traffic signs - if it is a speed limit sign the candidate has ignored). Not making proper use of the accelerator means just that; either accelerating too harshly, so that the car shoots off like a scalded cat when the clutch is let up, or not accelerating enough so that the car stalls and sits at the traffic lights for ages, while all around are shaking their fists and blowing their horns in the manner usually associated with the less patient members of the motoring public. This is not to say of course that if you stall the car on a driving test you will fail. Examiners are human and will take some account for the fact that you are nervous, and lack experience. But if you make a habit of it and make it obvious that you are unable to control the accelerator, then as I said earlier, perhaps you should not be there, and this is the view the examiner will take.

Proper use of clutch. From experience I would say that this is the most difficult item to teach the control of, and the most difficult one for learners to master. Some people (often middle aged women) never do master it and have to be advised to change to a car with automatic transmission, or give up. As far as failing the test on the incorrect use of this control goes, as I said before, stalling the car would not necessarily mean failure, but if it was obvious that you had not mastered clutch control it would. Stalling the car is of course caused by lack of control of the clutch or accelerator, or a combination of both.

Proper use of gears. This means selecting the correct gear for the road situation or gradient, and handling the gear lever sympathetically. Too many learners handle the gear lever like left handed gorillas trying to stir porridge. Most modern gearboxes have quite a smooth operation and do not require the superhuman effort that people seem to think is necessary.

One other point should be mentioned under this section. It is not necessary these days to go up through the gears or down through the gears. If for instance, you are approaching a set of traffic lights in fourth gear, and they change to amber, it is quite permissible to stop in fourth gear, then select first, ready to move off. However, most driving instructors still teach their pupils to go up and down through the gears using each and every gear. There is an excellent reason for this. The fact that modern gearboxes are capable of going at speed from fourth to second say, is irrelevant. Learners are generally taught to use all the gears because the correct selection of the appropriate gear in relation to road and engine speed is directly related to something which is very difficult to teach — anticipation, or reading the road ahead if you like. For example, approaching an empty pedestrian crossing where there are pedestrians on the pavement on both sides of the road. You will know, if you have read the Highway Code, that pedestrians do not actually have priority until they are on the road. However, most pedestrians, and in particular children see pedestrian crossings as magic carpets which will instantly whisk them safely to the other side once their big toe has made contact with the road surface. Bearing this in mind, most instructors realise that an empty crossing will not necessarily stay empty, and will teach their pupils to reduce speed as they approach. Obviously if you reduce speed, you should select a lower gear to suit the new speed. If a pedestrian then steps on to the road while you are still approaching you can change down again. What will usually happen then if you have approached correctly, is that the pedestrian will see you decreasing speed and will continue to cross, you will often find then that you can accelerate away without having to stop. If however, you consistently approach hazards in fourth gear like a horse with blinkers on, you will find yourself doing a fair bit of stopping and starting. Perhaps you won't fail for not making proper use of gears, but lack of anticipation and the consequential mis-use of the gears might help you to fail under another section.

Another fault which could be marked as incorrect use of the gears could be coasting, that is going along with the clutch down. You would probably get away with this once or twice, but if you did it going around a corner for instance and lost control you would certainly fail, (or crash!).

Make proper use of the gears

Proper use of footbrake. Obviously you should not brake too harshly, but on the other hand you need to brake enough to stop where you want to stop. It's a bit pointless braking so hard that you stop ten yards from a junction and are unable to see anything on the major road. By the same token it will not make you the most popular learner in town if you don't brake enough and the car hangs out of the junction halfway across the road before you have even had time to look.
Braking on bends remember is also not to be advised. Make sure you get your car to the correct speed for the bend before you get there, and accelerate to leave it. Don't leave things so late that you have to stamp on the brake like a demented grape crusher halfway round the bend. The footbrake is such a basic item in the control of the car that I must say once again, if you fail on this point, you should not be there.

Proper use of handbrake. I have often known candidates to fail on their mis-use of the handbrake who have been convinced that it was because they pulled the thing on without pressing in the button. True, the correct way is to press the button in and do it quietly, but this is only because of the wear that the ratchet receives, it has nothing to do with road safety. If leaving the button out and listening to the clicks makes you apply the handbrake adequately for a hill, that would be better than pressing the button in and only getting it half on. It may be quieter, but the car will make an awful noise when it runs into the vehicle parked behind you! Remember what I said earlier, faults will generally only be marked as serious or dangerous if they affect someone else. Making a noise with the handbrake will only affect the rachet, not other road users. Incidentally, I usually cover about 100,000 miles before I change my cars, and I let my pupils pull on the handbrake with the button out. I have never had a pupil fail for it, and I have never worn out a handbrake before wearing out the car! One other fault which could be serious; pulling the handbrake on before you have stopped when approaching a junction or pedestrian crossing may not fail you on its own, but remember your brake lights will not come on without the footbrake, and if the person behind runs into you or has to stop suddenly because of this you certainly would fail.

Proper use of steering. I have heard many candidates who have failed on this item say that it must have been because they crossed their hands. I have also heard instructors telling their pupils that they must have failed because they crossed their hands. You would be well advised not to cross your hands on the steering wheel, the correct method is taught for good reason. Crossing the hands can be dangerous. Having said that, an examiner would not fail you for it if you did it safely. If however, you were half-way round a corner and your hands collided with each other while you were in the process of crossing them, and you lost control, then of course you would fail; not for crossing your hands, but for what happened as a direct result of that.

Make proper use of steering.

Candidates who fail on steering do so because they mount the pavement, wander out into the middle of the road or neglect to straighten up after a turn thereby causing danger to other road users. All in all the sort of things associated with someone who has not had sufficient practice and has not yet mastered the steering. The sort of person who crosses their hands is the sort who has had years of practice, often on a farm, in a variety of vehicles and is quite competent. This is not to say that hand-crossing should be condoned, but it is safer than someone inexperienced who can not yet steer at all.

5. **Move away/safely/under control**

Moving away safely means just that; you must not cause danger or inconvenience to other road users. If you are in any doubt, then wait. Remember also that you will not see anything beside you by looking in your mirrors, only behind you. So before you move away check over your right shoulder. You will not be very popular if you move away gazing at your mirror while knocking a cyclist flying who happens to be in the process of overtaking you.

Move away safely.

Moving away under control. This of course refers to clutch control again. As I have said before, you will not necessarily fail if you stall while you are attempting to move away, but the examiner will know the difference between someone whose left leg develops a chronic case of the shudders because they are nervous, and someone who is not yet ready to be let loose unaccompanied.

6. **Stop the vehicle in an emergency/promptly/under control**

This is an exercise which I personally have always thought of as a bit of a daft thing to test people on; the reason being that it is very difficult to get learners to behave for test purposes in the same way that they would in a real emergency.

Most drivers, and especially newly qualified ones, when confronted with a child suddenly running out in the road in front of them will stamp on the footbrake with the force of a mechanical pile-driver, lock all four wheels and skid straight into either the very hazard they are trying to avoid or leave the road completely.

For the purposes of a driving test, most learners can be taught to brake quickly but gently so that the car comes to a stop without going out of control. But then this excercise is often done uphill, in third gear, at between 15 and 20 miles per hour. Somewhat different from say, 75 miles per hour on the M1 in pouring rain or even black ice.

I suppose the only way for the driving examiners to get some idea of how a candidate would react in a real emergency would be to have one arranged somewhere along the test route. Perhaps each test could be conducted by two examiners. One whose job it would be to sit in the car and give directions, while the other one would pop out from behind a parked vehicle and stand in front of the candidates car somewhere along the route. This would probably be a better system as the candidate, although expecting it at some time, would not know exactly when, and the examiner in the car would have a much better idea of how the person might react in a real emergency. The other examiner's opinion of course would be invalid as nine times out of ten he would de dead!
Still, perhaps this system should be considered as one of the fringe benefits would be to decrease the national dole queue because of the high wastage of examiners.

To be serious now, there is nothing difficult about stopping the vehicle in an emergency on a driving test. The examiner will give you ample warning, and as I said you will not be going very fast, and probably will not even be in fourth gear. If you are unable to satisfy the examiner on this point during a test, I shudder to think what you might do if confronted with a real emergency.

Stop the vehicle in an emergency under control.

7. **Reverse into a limited opening either to the right or left/under control/with due regard for other road users**

Learners have often said to me while struggling to reverse with some acceptable degree of accuracy, "I'll never do this again once I've passed!" They are missing the point of course. The reason for this exercise being part of the test, is so that the examiner can rest assured that if he releases you amongst the poor defenceless motoring public, you will not demolish half a dozen checkouts and their respective till assistants every time you attempt to reverse in or out of a space in Tesco's car park.

Bearing this in mind, it should now become obvious that you do not necessarily have to reverse perfectly in one movement and get it right first time. If you were reversing into a parking space, and you saw that you had mis-judged the angle and were on a collision course with a parked vehicle, you would hardly keep going and hit it thinking that you only had once chance. You would probably stop, go forward, straighten up, and come back again. Two or three times if necessary. That is exactly what you should do on a driving test if you see it going wrong. The examiner only wants to see that you are capable of driving the car backwards and putting it where you want it, safely.

Having said all this, it would be a total waste of time to wait until you have mounted the pavement before having another attempt. The examiner will have to assume that you are quite happy to collide with whatever other vehicles may have the mis-fortune of being parked near where you manoeuvre. You will see that you are getting too close to the kerb before you hit it, stop the car and put it right. Don't even bother to ask permission. The examiner will only tell you to do what your instructor would expect.

Notice that there are two main points to this exercise. "Under control", and "with due regard for other road users". To deal with the control first, this is obvious. You must control the clutch so that the car goes backwards at a reasonable rate and you are able to position it properly. It's no good bombing backwards like a rocket completely out of control.

To deal now with the matter of having due regard for other road users. You must not cause danger or inconvenience to other road users. There is no point reversing accurately with perfect clutch control if you completely disregard the mayhem you may be causing by not taking notice of what is going on around you. For instance, remember that the front of your car will swing out in the road as you turn, so watch for traffic approaching from the right and stop if you see anyone. Remember also that if a vehicle tries to emerge from the road you are reversing into it will be difficult for the driver to see safely if you are stuck on the corner, so go forward again out of the way, and start again when it is safe.

"Other road users" of course, means anyone who happens to be using the road, not only other drivers. Watch out for pedestrians, especially children. It may be difficult for you to see them behind your car.

One final point. If you are driving a van or similar vehicle with limited visibility to the left, or a left hand drive vehicle, you will be asked if you would prefer to reverse into an opening on the right.

The correct method for executing this manoeuvre is comprehensively explained in the Department of Transport Manual "Driving". Just make sure that you don't go back with your head hung out of the window gazing at the pavement and completely ignoring anything that may be behind you or coming towards you, and remember to reverse far enough so that you can *safely* cross over to the left *before* emerging from the junction you have just reversed into.

Reverse into a limited opening on the right with due regard for other road users.

8. **Turn around by means of forward and reverse gears/under control/with due regard for other road users.**

This exercise often brings the same sort of comment from learners as the reversing exercise, in that they do not intend to ever do it again once they have passed their test. Of course, as with the reversing exercise, it is done simply to show the examiner that the candidate has control of the vehicle. Not only control of course, but notice the familiar old phrase, "with due regard for other road users".

To deal with the control aspect first, this manoeuvre is more commonly known as the three point turn. Most driving instructors tend to get quite irate with anyone who refers to it as a three point turn, (usually other instructors), but they insist on their pupils doing it in three movements. Personally I could not care less what it is called, as long as it is done properly.

A driving examiner will not necessarily expect you to complete the exercise in three movements, depending of course on why you are unable to. It will either be because you are driving a large car or van, or because you have been asked to do it in a fairly narrow road. There is nothing at all wrong with doing it in four or five movements in these instances. If however, you are unable to turn a mini around in a road the size of an airstrip in less than half-a-dozen shunts, then the examiner will have to decide whether or not you really have control. The less control you have, the more likely you are to receive a serious fault.

To deal now with due regard for other road users. The same rules apply as with the reversing exercise, you must not cause danger or inconvenience to other road users. There is a very contentious point amongst driving examiners about the way a learner should handle approaching traffic while executing this manoeuvre during a driving test. I have heard two opposite views from two different supervising examiners who both insist that they are quoting Department of Transport policy. The question is, if a vehicle approaches from either direction while you are in the middle of the exercise, should you wave them on? Some examiners think that you should, others do not agree.

The safest way by far is for you to be guided by whatever the approaching driver does. If the approaching vehicle stops and the road is otherwise clear, then you should go. There is no point in two drivers sitting looking at each other, eventually you will probably get fed up and go together. If, on the other hand, the approaching vehicle keeps going then you should wait. If you remember this, and just do the opposite to what the other driver does, it will be quite safe, and there will be no need for you to give signals to anyone.

Finally, on the subject of observation, a very common fault amongst learners when turning the car around is that they furtively look left and right over and over again as if they are at the centre court at Wimbledon, but never once look behind them. Quite amazing when you consider that that is the way the car is going. You would hardly drive forwards looking behind you, why drive backwards looking forward?

Turn round by means of forward and reverse gears — with due regard for other road users.

9. **Make effective use of mirror(s) well before/signalling/ changing direction/slowing down or stopping**

This is probably one of the most contentious items on the failure sheet. I have lost count of the number of times I have heard people say that they failed their driving test because they didn't look in their mirror often enough. The most common advice you will get on this subject from friends, relatives, the next door neighbour and all manner of self appointed experts who don't really know an awful lot about the subject, is that you should put the mirror at an angle so that you have to move your head to see in it. The stupid blind examiner will certainly notice then every time you look! One of the problems with this theory is that the odd time you do forget to look will stand out like a sore thumb.

Other well meaning advisers have been known to tell learners that they will be o.k. as long as they check the rear view mirror every ten seconds/five seconds/two seconds, depending on who is giving the advice. Perhaps the most useless piece of advice I personally have ever heard came from the father of one of my pupils, who apparently told his son to look in his mirror every third lamp post! This only came to my notice when the learner in question seemed to be aiming at other vehicles and pedestrians when we were on a road with lamp posts, but where there were no lamp posts drove reasonably well. Except of course that he didn't bother with the mirror! Once we had managed to sort out this minor problem and got him to concentrate on the road and other road users rather than counting lamp posts, his driving improved. The wording of this item on the failure form is **"Make EFFECTIVE use of mirror(s)".** There is absolutely no point at all in gazing in your mirror just before you are about to change direction for instance to position your car for a right turn, if you completely ignore the fact that a motorcyclist say, is in the process of overtaking you, and you knock him flying. No one would dispute the fact that you looked in your mirror, but you must take the appropriate action according to what you see. That is making **effective** use of the mirror.

One of the common failings at least in the test area that I use, is when the examiner and candidate are going along a dual carriageway and the candidate is asked to turn right at the roundabout ahead. This of course means changing lanes to the right normally when much faster traffic is overtaking. Many candidates mis-judge the speed or distance of overtaking traffic, cause danger and inconvenience to other road users and fail for not making **effective** use of the mirror before changing direction. It would be much better if you think it is unsafe to change lanes to go straight ahead or turn left, the examiner will not be happy about having his route plan messed up but he cannot fail you for going the wrong way. (As long as you don't turn left indicating right!).

Remember, it is a test of your driving ability, not your hearing. Another very frequent cause of failure is if the candidate after being asked to "follow the road ahead" at a roundabout cuts across from left to right and back again whilst driving through the roundabout. If any other vehicle is inconvenienced the candidate will certainly fail for not making **effective** use of mirror.

Of course when candidates return to the test centre having failed for not making effective use of the mirror, they insisit that they looked in the mirror frequently. No one is suggesting that they didn't, but they didn't take the appropriate action.

Make effective use of mirror(s) well before changing direction

10. **Give signals/where necessary/correctly/in good time**

Where or when are signals necessary? Quite simply if another road user would benefit from a signal then you should give one. This does not mean that you should use your indicators every single time you move away from the side of the road, or stop on the left. Nor should you go around like a flashing beacon past every single parked vehicle that you see, and indicating back in again after you have passed them.

I remember being stood outside the driving test centre once when a young lady was about to move away at the start of her test. The indicators went on almost immediately after she had started the engine, she certainly did not have time to look to see if it was necessary to signal. A bus was coming along behind, and the driver, seeing the candidate signal, decided to stop to let her out. However, she was not aware of this act of generosity because she had not yet started looking, she was still adjusting her seat, putting the seat belt on and tidying her hair. By the time that she became aware of the presence of the bus, and that the driver was waiting for her to go, the bus driver had got fed up with his chivalrous gesture being ignored and had decided to get moving again himself; at the very same time that the young lady finally decided to move. There was then a deafening screech of tyres from the bus as the driver braked hard, and an even more deafening sound as he blew his horn. I never did know the result of that test, but if she did fail on that point she managed to fail before she even got moving.

Give signals where **necessary**. Before you move off, have a look. Is there anyone who may benefit from a signal? Other people perhaps sat in stationary vehicles, they may also be thinking of moving off; pedestrians about to cross the road; vehicles coming towards you, there may be plenty of room for both of you, but the oncoming driver may be about to take up more room than he needs. Probably the only person who would **not** benefit from a signal would be someone coming from behind, all that will achieve is confusion. There is one exception. If you want to move away on a busy road where the traffic is continuous, say in the rush hour for example.
If you are not indicating no one will know that you want to pull out so no one will give way to you. If you are indicating sooner or later there will be a gap and someone will probably let you in to it.

Indicating around stationary vehicles. It is not necessary to indicate around everything. If you are overtaking a cyclist for instance and the drivers behind are unable to see it because you are blocking their view, then they need to know that you are moving out. If however you are overtaking parked vehicles it's a bit daft indicating in and out when it is quite safe to stay out and overtake the lot. You should never indicate back in after overtaking a vehicle anyway, it just gives the impression that you are parking.

Give signals **correctly**. Obviously don't signal left if you are turning right! Another thing you could be faulted for under this section is not cancelling the indicator after you have finished with it. Failing to cancel the left indicator when pulling up for instance, and then driving away still indicating left would be marked under this section — it would be a minor mark unless it was persistant, or it involved a passing vehicle.

Don't forget the signals at roundabouts either. Many people when going straight ahead at a roundabout approach correctly in the left hand lane, but then for some reason signal right as they enter the roundabout. This could be a serious fault if it affected someone else in the right hand lane turning right because they might think that you are about to come across in front of them.

Give signals correctly

Give signals **in good time**. This is obvious. It's not much use signalling at all if you do it too late to be of any practical use to anyone.

11. **Take prompt and appropriate action on all/traffic signs/road markings/traffic lights/signals given by traffic controllers/other road users**

Traffic Signs. Remember the familiar old phrase, you must not cause danger or inconvenience to other road users. You must keep your eyes open and notice road signs, they are there for your benefit.

Stop signs for instance at road junctions, it's no good treating the junction as if it has a give way sign and creeping slowly out. Creeping is not stopping.

One important point to consider in this section is that the examiner at the start of the test will tell you to follow the road ahead throughout. Obviously if you are approaching a no right turn sign he will not expect you to ask him which way he wants you to turn. You must look for traffic signs and act on them, the examiner is not there to look for you.

Take prompt and appropriate action on all traffic signs.

Road markings. The only difference between road markings and traffic signs is that the road markings appear only on the road surface. Once again, you will be expected to notice road markings and act on them. Here are a few examples of things you should **not** do. Overtaking a slow moving vehicle while there is a solid white line nearest to you in the centre of the road, blocking a box junction, approaching a roundabout in the left hand lane to go straight ahead when there are arrows on the road telling you that the left hand lane is for turning left only, blocking entrances marked with "keep clear" road markings and driving along in a lane with a turn left or turn right arrow when you are meant to be following the road ahead. (The one with the straight arrow).

You will remember back at the start of the book, I pointed out that although driving examiners do not usually fail people on their lack of knowledge of the highway code, it would be a good idea to learn it so that your lack of knowledge does not help you to do something seriously wrong during the test?

A good example could be the way you might behave on a box junction. For those who have not yet read the highway code and for those who have read it but don't understand it, the section dealing with box junctions says, "you must not enter the box if your exit road or lane from it is not clear. But you may enter the box when you want to turn right and are prevented from doing so only by oncoming traffic or by vehicles waiting to make a right turn". Clear? No, I didn't think so. Nor is it for most people. The only bit most learners remember is that they must not enter the box unless their exit is clear. (I say learners, because qualified drivers do not even seem to **see** box junctions). So when the examiner has asked you to turn right and you are sat **outside** a box junction giving way to oncoming traffic, don't be surprised if you fail for not taking prompt and appropriate action on road markings.

Take prompt and appropriate action on all road markings.

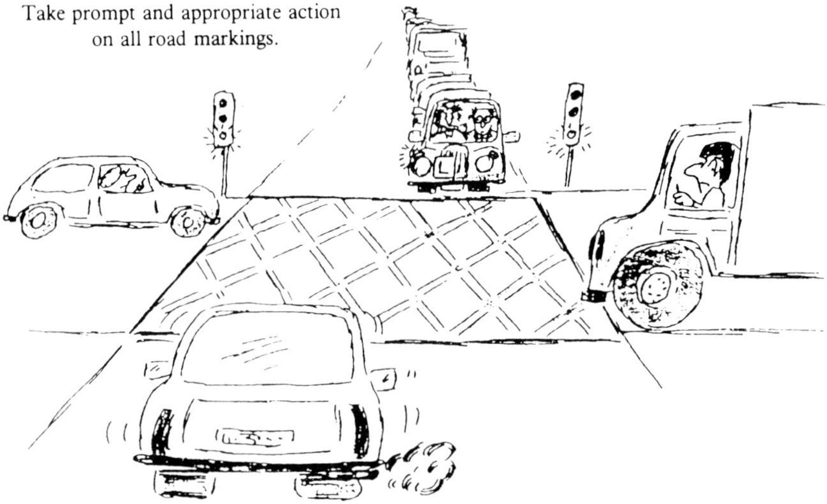

Traffic Lights. Quite self-explanatory really!

Take prompt and appropriate action on all traffic lights.

Signals given by traffic controllers/other road users. Traffic controllers are traffic wardens and police officers, and any signals you receive from them will be quite clear and should present no problems. Signals by other road users however, may present problems as the people giving them will generally speaking not be people authorised to control traffic, which in most cases is probably just as well.

Road users remember could be pedestrians as well as drivers. We have all seen the little old ladies waving cars out of road junctions telling the drivers that the road is clear. Of course their idea of "clear" could be "there's a forty ton lorry hurtling towards you at ninety, but he's twenty yards away so if you put your foot down you will make it!". Then again we are all familiar with flashing headlamps.

This well known signal of course means, "I am giving way to you" or "stay where you are I'm coming through", or "Hi Fred! I haven't seen you for ages", or "damn! that was meant to be the windscreen wiper!" and last but not least, "you can go first, providing you get moving in the next two seconds or preferably less! "

This last signal is mostly intended for learners, because as everyone knows, learners move off quickly and efficiently, and generally with far more skill than experienced drivers.

Take prompt and appropriate action
on all signals given by other road users.

To be serious now, you are the one behind the wheel, you must decide whether or not it is safe to proceed. Imagine for instance, you are waiting in a busy road to turn right into a side road, and a driver coming towards you flashes his headlamps. Nine times out of ten he will be giving way to you. But it would be wrong for you to go immediately while the oncoming driver is travelling at speed, he may not stop. On the other hand it would also be wrong for you to wait **after** the point when it becomes obvious that he **is** giving way to you. This would cause confusion. Remember, you are responsible for making sure that it is safe.
Think for example of the possibility of an overtaking cyclist on the inside of the car that is giving way, because I can assure you, the other driver will not think about it.

12. **Exercise proper care in the use of speed**

If you see this item marked on your failure form it should be pretty clear to you that you went too fast at some time during the test, or perhaps even all through the test. But how fast is too fast? Obviously, if you break the speed limit you stand an excellent chance of failing, but failing on this point may not necessarily mean that you did exceed the limit. For example 30 mph would be much too fast if you were driving in a busy town centre and there was a whole army of pedestrians crossing the road in front of you. You must exercise proper **care** in the use of speed; which of course does not mean that you should not **use** speed, just that you should use it in the right place at the right time.

Exercise proper care in the use of speed.

13. **Make progress by driving at a speed appropriate to the road and traffic conditions/avoiding undue hesitancy**

This section of course goes hand in hand with the last one. While you are exercising proper care in the use of speed, don't be so petrified of going too fast that you end up plodding along at four miles a fortnight making everyone else wish that learners were only allowed to use the roads on Sundays. Having said that, I know from experience that you would have to drive extremely slowly to fail on this item.
However, don't forget that it does not only refer to driving slowly, you can also be faulted for waiting too long in traffic at a junction.

Generally speaking, if you are not sure whether to go or not, then it is probably better not to. On the other hand, in modern traffic conditions in most towns or cities you would be unlikely to be fortunate enough to see a completely clear road no matter how long you were to wait at a busy junction.
You have to decide whether or not a gap is big enough for you to jump into, and whether or not the approaching traffic is going slowly enough for you to have time. Remember, you must not cause danger or inconvenience to other road users. If you are unable to make these judgements accurately, then perhaps you need more practice. Item 11 can often be associated with this item on the failure sheet, (take prompt and appropriate action etc). "Prompt" of course means without taking so long to decide that everyone else gets frustrated and general confusion is the order of the day.

You will be expected to know what the speed limit is on the road you are driving on. You may have passed a national speed limit sign say, but you did not notice it so you think just to be on the side you will drive along at 29 miles per hour, just in case you are in a 30 mile limit. You will probably not get away with it, especially if you are holding up other traffic.

Make progress by driving at a speed appropriate to the road and traffic conditions.

14. **Act properly at road junctions: regulate speed correctly on approach/take effective observation before emerging/ position the vehicle correctly - before turning right, before turning left/avoid cutting right hand corners.**

From my own personal experience I would say that driving test candidates fail on road junctions more frequently than on any other item on the failure form.
Before I go on to deal with the subsequent sections of this item in detail, just a quick reminder about what a road junction is. Most candidates tend to look at this item on the failure form and think only about 'T' junctions, where a minor road meets a major road. "Road junctions", for the purposes of the failure form, are 'T' junctions, uncontrolled cross roads, cross roads controlled by traffic lights or road markings, roundabouts and mini-roundabouts.

Regulate speed correctly on approach. This means exactly what it says. You should not race up to road junctions at breakneck speed. Slow down gradually as you approach, and select the appropriate gear for your speed so that you are ready to go if the road is clear or stop if it is not.
Take effective observation before emerging. Many failed candidates when seeing this item marked, insist that they looked at every junction they came to. There are two major points to consider here though; firstly, they probably did not look at the junctions that they didn't know were there, and secondly, on the junctions that they did see, did they take **effective** observation? What sort of twit would not notice a junction you may ask? Quite a few actually, when you consider that the sort of junctions learners often fly across without looking, are uncontrolled cross roads and mini-roundabouts. For some reason they just don't see them, but then this is a disease not only suffered by learners. The big problems come of course when two drivers with the same eyesight deficiency are approaching the same hazard from opposite directions. Fortunately for the rest of us though, this is one of the rare sort of accidents which only affect the people who cause them. Anyone who **is** taking **effective** observation will avoid them.

Act properly at road junctions — regulate
speed correctly on approach.

Which brings me to the difference between observation and **effective** observation. On many occasions I have seen learners pull up at a road junction, have a look at the cyclist approaching from the right and pull straight out in front of him. It's a fat lot of good looking when you arrive at a junction if you don't take any notice of what you see. If that sort of thing happened on a driving test you could expect to be failed for not taking effective observation before emerging. The examiner is not necessarily saying that you did not look, just that you did not take the appropriate action on what you saw.

Act properly at road junctions — take effective observation before emerging.

One final point to consider. If you should manage to ignore either a mini-roundabout or an uncontrolled cross roads during a driving test you will normally see "regulate speed correctly on approach" AND "take effective observation before emerging" BOTH underlined on the DL 24. Obviously, you did not regulate your speed approaching the junction because you did not see it.

Position the vehicle correctly/before turning right/before turning left

A common fault among most drivers, (not only learners) is that when they want to turn right they tend to position themselves either on the right hand side of the road, or if not completely on the right, more to the right than the left. I understand the reasoning behind this, they are trying to pick a point "just left of centre". This is fine, if the road is wide enough to justify it. If however, the road is not wide enough for two vehicles on the left hand side, there is no point at all in moving more to the right than your normal driving position.
You will only succeed in making it difficult for drivers who wish to use the other side of the road coming towards you. Also, if you take up too much room when you are emerging from a road junction you will simply hamper vehicles turning into the road. There is one other mistake that this item on the failure form could be referring to, and that is staying on the left hand side of the road when turning right in a one-way street. A fault marked under "position before turning left" would mean that you pulled out too wide on the approach to a left turn.

Act properly at road junctions — position the vehicle correctly before turning right.

Avoid cutting right hand corners. Drivers generally, not only learners, first of all position themselves too far to the right as they approach a right turn, and then turn too early cutting the corner. Many drivers tend to make this mistake either near their home, or near their place of work, or some other place where they drive frequently. They get so accustomed to a certain road being clear as they drive into it, that they assume it will always be clear. When they do finally meet something coming towards them they usually look totally bewildered as if the other driver has no right to even be on the same planet, let alone using the same road!

It is possible to get away with this sort of behaviour during a driving test, depending of course on how blatant it is, and whether or not you cause danger or inconvenience to other road users.
If you see well ahead that the road is clear and you cut the corner, you may get away with it. (Once or twice). However, the examiner will soon notice if you do it on every junction whether it is clear or not. The fact that it did not affect anyone else during the test could just be a matter or luck. If on the other hand you cut a right hand corner and you **do** cause danger or inconvenience etc, you will certainly fail.

Perhaps it would be better to do it properly every time. It could be a lot less effort than trying to decide whether or not it's safe to do it incorrectly.

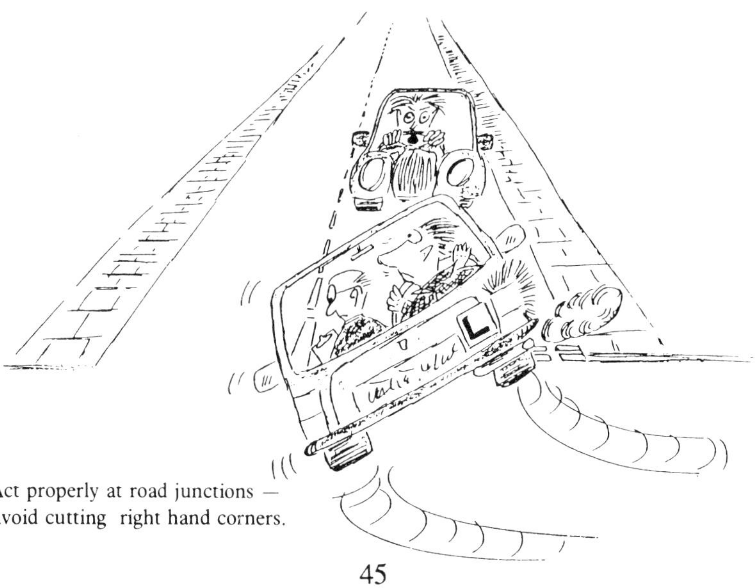

Act properly at road junctions — avoid cutting right hand corners.

15. Overtake/meet/cross the path of/other vehicles safely

The majority of test candidates don't seem to understand this section when they see it marked. The bit about overtaking does not seem to present any problems, most learners remember after a test if they overtook something dangerously. Due to the low level of speed on driving tests this usually refers to cyclists, milk floats, buses stopping at bus stops and parked vehicles or other obstructions.

Meet other vehicles/cross the path of other vehicles safely. Both these items seem to cause confusion. Meeting other vehicles generally relates to the situation where you are driving along a road with vehicles parked on one or both sides and you meet other traffic coming towards you. Crossing the path of other vehicles would relate to a situation where you intend to turn right, and it is necessary to give way to on-coming vehicles.

To deal with meeting other vehicles first, if there are parked vehicles on your side of the road only, on-coming traffic has priority. If there are parked vehicles on both sides of the road, the drivers going downhill **should** give way to those coming up. However, if the road is level, common sense must prevail. Don't make the mistake of thinking that on-coming vehicles will always give way to you when they have stationary vehicles on their side of the road. Forcing another driver to stop because you are being bloody minded would be a serious fault, regardless of who has priority.

Meet the path of other vehicles safely.

Crossing the path of other vehicles. If you are turning right you must give way to on-coming vehicles. If you are in doubt about their speed and distance, wait. There are no prizes for the first one back to the test centre.
There is one very common fault among learners when crossing the path of other vehicles. This happens when waiting to turn right at a cross roads controlled by traffic lights. When the lights change to green, they shoot off round the corner like something possessed completely disregarding the oncoming traffic, which, of course, also have a green light! Remember, a green light does **not** mean that you have priority.

16. **Position the vehicle correctly during normal driving**

To return to what the examiner says when he first gets into the car. "Follow the road ahead unless road markings tell you otherwise, or I ask you to turn". This means that usually, you will be driving straight ahead keeping to your left. Driving in the gutter or out in the middle of the road would both be obvious faults under this section, but less obvious ones might be driving along the right hand lane of a one-way-street or a dual carriageway when the examiner has not asked you to turn right, and there are no road markings telling you to drive on the right.Of course, there is nothing wrong with using the right hand lane of a one-way-street or a dual carriageway, but by the same token, if you have no reason to be in that lane, you should keep to your left. This seems to be a very difficult thing to get across to learners. I realise that after you have passed your test you will generally know where you are going and will be in the appropriate lane without being told, but try to bear with driving examiners, and for that matter, with us poor instructors! None of us wants to keep chirping "keep to the left!" "keep to the left!" like parrots with verbal diarrhoea, so if you always keep to the left unless there is some reason not to, it won't be necessary.
I have lost count of the number of times learners have said to me after turning a corner, "which lane do you want me in?".

Position the vehicle correctly during
normal driving.

17. **Allow adequate clearance to stationary vehicles**

Driving test candidates often fail on this point while they are busy failing for not meeting other vehicles safely.
What happens is, they drive along a road crammed with stationary vehicles and see a vehicle coming towards them. They quickly assess that there is room for them to "squeeze" through, and this is exactly what they do, not leaving room for a sliding door to open, never mind one that opens in the conventional manner.

Quite a number of times I remember telling learners to move out and leave room for a door to open, and they have said "why? there's no-one in the car!"

I will always remember one of my pupils saying just that to me as we were about to slip past a stationary car leaving just about enough clearance for a well worn feeler gauge lubricated with butter, when the driver's door flew open to reveal the upside-down face of a dirty D.I.Y enthusiast who looked like he had just spent the last two days on the floor of his car trying to fit a radio. Realising that he probably had no wish to spend the next two days fitting a new door, I trod on the dual brake; whereupon the driver behind in a mini who had spent the last mile or so behaving as if he was attached to my tow-hook, actually became attached to it quite firmly. So much so in fact that when we finally managed to separate our cars, my car was still in possession of his radiator!

"Adequate" clearance means just that, not two inches, but ample room for a door to open. You never know when it might.

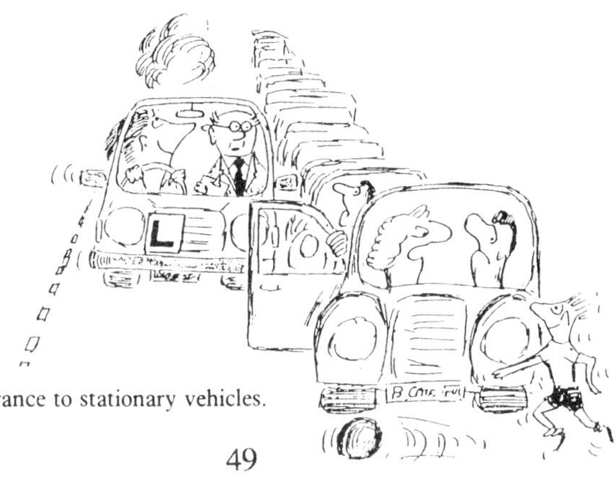

Allow adequate clearance to stationary vehicles.

18. **Take appropriate action at pedestrian crossings**

This section also relates to pelican crossings remember. These crossings cause more arguments between drivers and pedestrians than anything else, simply because pedestrians usually disregard the instruction "do not start to cross" when the green man is flashing. Which is why, when pedestrians see the flashing green man, and drivers see the flashing amber light, the Highway Code tells drivers to "give way to pedestrians". It does not say "put your foot down and make the stupid pedestrians run because they should not have started crossing anyway!"

To turn now to ordinary pedestrian crossings, or zebra crossings as they are often known. The Highway Code tells pedestrians that drivers do not have to stop for them until they have started to cross, which probably accounts for so many pedestrians treating crossings like magic carpets.

Always approach crossings carefully, watch the pedestrians either side walking towards the crossing, they may turn and suddenly walk to it. On the other hand don't do an emergency stop because a pedestrian walks up to a crossing and stands there looking at you. If they are stood waiting as you approach, then you have plenty of time to stop carefully and smoothly.

If you arrive together and they wait, you might as well carry on, it's better than stopping suddenly.

One final point. **Never** wave pedestrians across the road anywhere, (not only pedestrian crossings). There is nothing polite or considerate about this action, it is down-right dangerous.

You have no control at all over the actions of other drivers, it is not your job to tell pedestrians that it is safe, it is up to them to make sure that it is before they start to cross.

Take appropriate action at pedestrian crossings.

19. **Select a safe position for normal stops**

This book has not been written with the intention of replacing the Highway Code, so I do not intend to waste time listing all the places that are quite clearly shown in that booklet as places where you should not stop.

It is not the examiners job to tell you **exactly** where to stop, it is your responsibility to make sure that you pick somewhere safe. On occasions throughout the test the examiner may ask you to stop in a certain place, for instance behind a staionary vehicle leaving room to pull out. But sometimes he will ask you to stop "in a convenient place on the left". **You** have to pick the spot and decide whether or not it is safe.

Select a safe position for normal stops.

20. **Show awareness and anticipation of the actions of/ pedestrians/cyclists/drivers**

This is one of the more contentious items on the failure sheet, because it is quite easy for a test candidate to fail the driving test through someone else's bad driving, or the careless and thoughtless acts of pedestrians or cyclists. Then again not all candidates fail on this item through someone else's stupidity, often it can be because they did not anticipate a move or action which may be a perfectly correct one.

I have always believed that accidents are caused by two or more people. There is the person who sets the accident situation up in the first place by doing something wrong, then there is the person who lets it happen.
For example: a driver going across a red traffic light and other drivers coming from the opposite direction not looking both ways, a cyclist overbalancing and falling off his bike in the path of a vehicle overtaking too close or a pedestrian walking out in front of a car when the driver did not even see him walking towards the edge of the pavement.

One sure way of getting marked for lack of anticipation is to drive close up to the rear of the vehicle ahead in traffic hold ups. Learners should allow enough room to be able to drive around the car in front — it may not move when it should.

As a rule, good drivers are never taken by surprise by anything. If you are turning into a one-way-street expect to see vehicles coming out of it, if you are driving in fog, expect to find a vehicle coming towards you with no lights, if you see a cyclist, expect him to fall off. No one expects you to drive with one hand on the wheel, and the other on your crystal ball, but most situations are quite predictable, be ready for them.

Show awareness and anticipation of the
actions of: other road users.

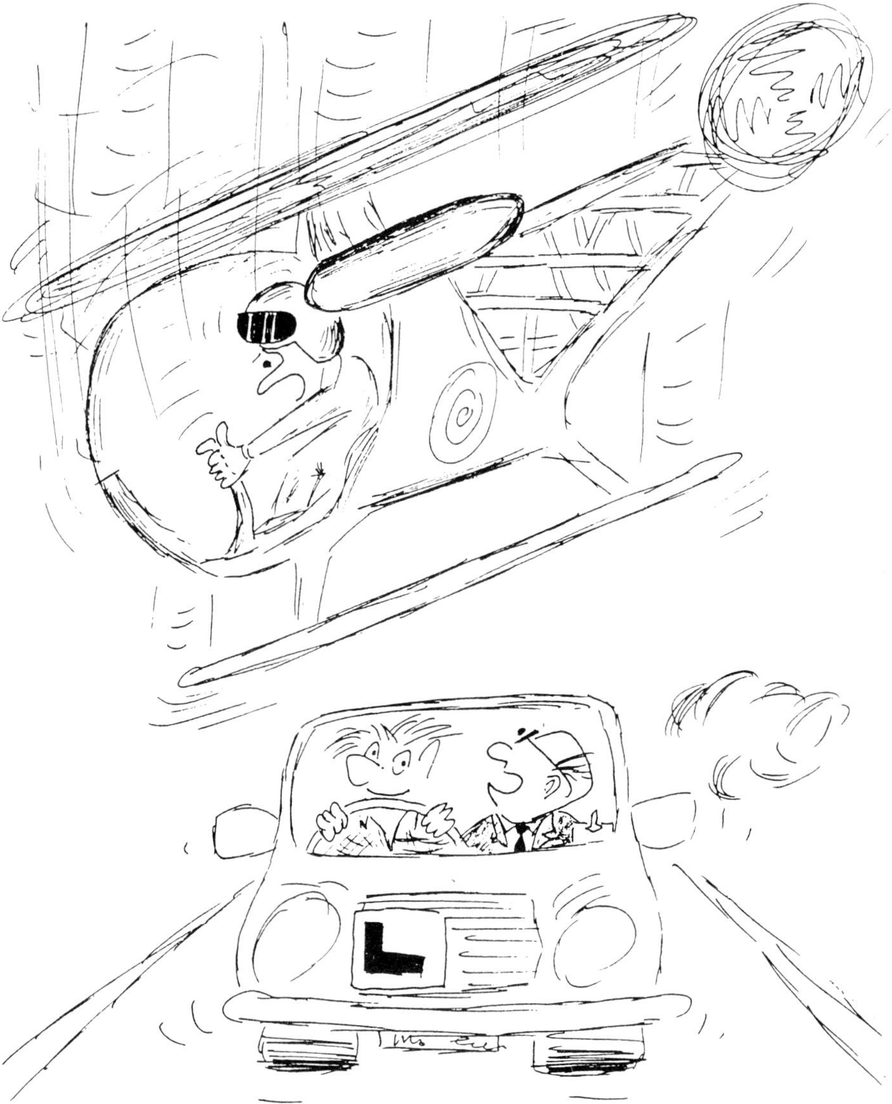

Conclusion

If you have just struggled through this book, you probably have reached the conclusion that I have a very light-hearted view of driving examiners, instructors, the driving test, learner drivers and road safety in general.

Nothing could be further from the truth, but you could be forgiven for thinking it. I have approached this very serious subject in a light-hearted manner because I personally know only too well how easy it is to become bored if something does not keep my attention.

I therefore set out to make this book as **readable** as possible, knowing how boring the Highway Code and the Driving Manual for instance are. Having said that, if you are a learner, you do really need to read these publications, but I do appreciate the problem, it is difficult to get interested.

A final word about the driving test now. Forget all the rubbish your friends and relatives tell you, they probably know very little. It is a sad fact about human nature that when people know very little about a subject, they make up the rest to fill in the gaps.

Driving examiners would rather pass you than fail you, but you have to put some effort in yourself. You won't be failed because it's a Friday afternoon, or because you have a Mohican hair-do, or because you pull the handbrake on without pressing the button in, or because sometimes you look in your mirror **after** you signal instead of **before**.

By the same token if you are an appalling driver, you won't be passed because you have a nurses uniform on, or because you say sorry every time you nearly have a collision or because you call the examiner sir! The examiner will only concern himself with the answer, as he sees it, to one question. Whether or not you are safe to be let loose unaccompanied.

Never forget, they too are human. If I was a driving examiner, I'm damned if I would let someone loose knowing that they might mount a pavement driving along and kill one of my children. Don't expect driving examiners to feel any different.

One last word about your driving lessons. This book is not designed to teach you how to drive. No book ever could. I have seen many books advertised boasting about the number of days the reader could learn to drive in if they were to fork out and buy a copy.

All instructors have their own methods, it is not for me to criticise. If your instructor recommends a certain book (other than the "Highway Code" or "Driving") then read it if you want to, and if you want to buy it. There is one thing though , I have learnt in the last seventeen years, and this view seems to be shared by other driving instructors of similar or more experience. That is, that nothing beats practice on the road, you can watch films, read books and sit in lay-bys talking theory until it comes out of your ears, but only one thing will ever stand a snowball's chance in hell of changing you from a ham-fisted, fumbling novice into a safe and competent driver - practice! Lots and lots of it. Get behind the wheel and drive as much as you can.

I hope you have enjoyed this book, it probably has only just scratched the surface of a very complex subject, but it may have opened your eyes a little to some of the 'trade' secrets!

LEARN TO DRIVE IN TEN MINUTES
THE END

Published by:-
Iron Box Publications
Iron Box Cottage, Whitecroft,
Dilton Marsh, Westbury
Wiltshire BA13 4DJ

Typeset and printed by:-
Kall Kwik Printing, Salisbury, Wiltshire